APR 22

In the Know

INFLUENCERS AND TRENDS

TECH

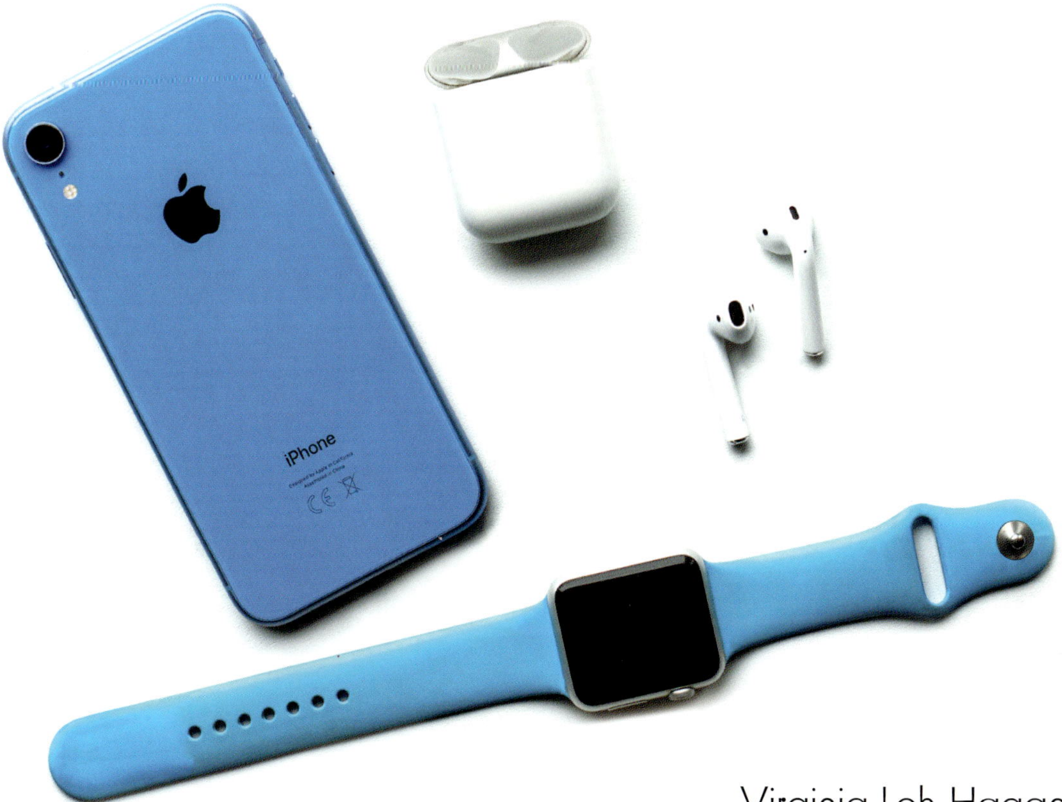

Virginia Loh-Hagan

45TH PARALLEL PRESS

Published in the United States of America by Cherry Lake Publishing Group
Ann Arbor, Michigan
www.cherrylakepublishing.com

Reading Adviser: Marla Conn, MS, Ed., Literacy specialist, Read-Ability, Inc.
Book Designer: Felicia Macheske

Photo Credits: © Christine Glade/Shutterstock, cover; © Daria Photostock/Shutterstock, 1; © DisobeyArt/Shutterstock, 4; © Pakhnyushchy/Shutterstock, 6; © Public Domain/Historic Computer Images/Wikimedia, 7; © Cineberg/Shutterstock, 8; © HJBC/Shutterstock, 10; © Sebastien DURAND/Shutterstock, 12; © By sklyareek/Shutterstock, 14; © Creativa Images/Shutterstock, 15; © PhotoColor/Wikimedia, 16; © insta_photos/Shutterstock, 18; © MatejTU/Wikimedia, 20; © gmn/Shutterstock, 22; © Max4e Photo/Shutterstock, 23; © NTNU–The Norwegian University of Science and Technology/Flickr, 24; © Aitor Serra Martin/Shutterstock, 26; © Artorn Thongtukit/Shutterstock, 28; © Hethers/Shutterstock, 31

Copyright © 2021 by Cherry Lake Publishing Group

All rights reserved. No part of this book may be reproduced or utilized in any form or by any means without written permission from the publisher.

45th Parallel Press is an imprint of Cherry Lake Publishing Group.

Library of Congress Cataloging-in-Publication Data

Names: Loh-Hagan, Virginia, author.
Title: Tech / by Virginia Loh-Hagan.
Description: Ann Arbor, Michigan : Cherry Lake Publishing, [2021] | Series: In the know: influencers and trends | Includes bibliographical references and index. | Summary: "We're spilling the tea on the latest trends. Curious about the lives of influencers? Want to know more about 5G technology or artificial intelligence? Read more to be in the know. This high-interest series is written at a low readability to aid struggling readers. Each book includes educational sidebars, throwback biographies of "OG" influencers, fast facts, and social media challenges, as well as a table of contents, glossary of keywords, index, and author biography"— Provided by publisher.
Identifiers: LCCN 2020032457 (print) | LCCN 2020032458 (ebook)
| ISBN 9781534180383 (hardcover) | ISBN 9781534182097 (paperback)
| ISBN 9781534181397 (pdf) | ISBN 9781534183100 (ebook)
Subjects: LCSH: Technology—Juvenile literature. | Technological innovations—Juvenile literature.
Classification: LCC T48 .L643 2021 (print) | LCC T48 (ebook) | DDC 600—dc23
LC record available at https://lccn.loc.gov/2020032457
LC ebook record available at https://lccn.loc.gov/2020032458

Cherry Lake Publishing Group would like to acknowledge the work of the Partnership for 21st Century Learning, a Network of Battelle for Kids. Please visit http://www.battelleforkids.org/networks/p21 for more information.

Printed in the United States of America
Corporate Graphics

Dr. Virginia Loh-Hagan is an author, university professor, and former classroom teacher. She has a smart home. She has many connected devices. She lives in San Diego with her very tall husband and very naughty dogs. To learn more about her, visit www.virginialoh.com.

Introduction | Page 5

Chapter **1** | **The Internet of Things** | Page 9

Chapter **2** | **Self-Driving Cars** | Page 13

Chapter **3** | **Artificial Intelligence** | Page 17

Chapter **4** | **Bionics** | Page 21

Chapter **5** | **Gene Editing** | Page 25

Chapter **6** | **Sci-Fi Reality** | Page 29

Glossary | Page 32

Learn More | Page 32

Index | Page 32

Influencers work with companies.
They help promote different products.

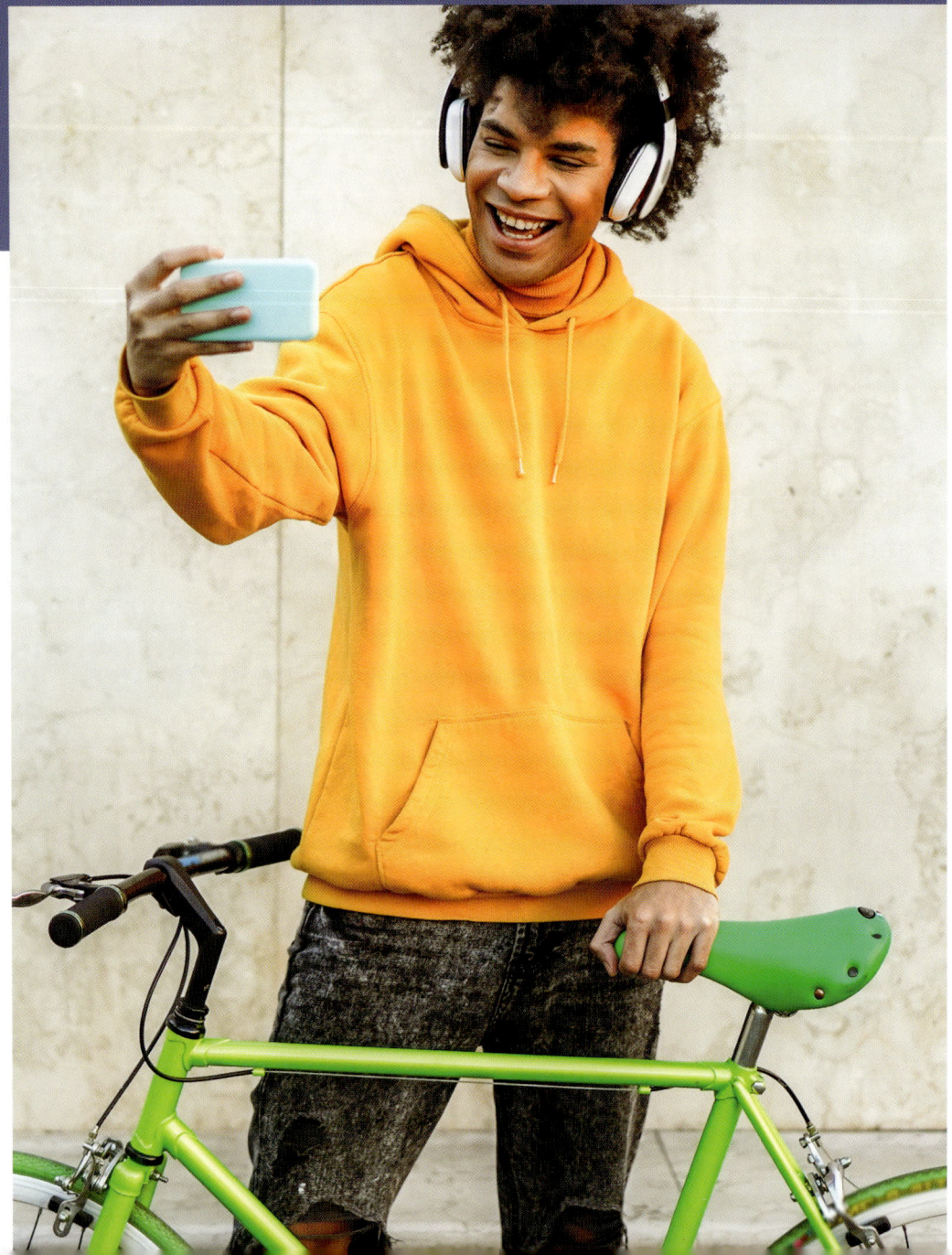

Introduction

What's **trending**? What's in? What's not? Trends are popular fads. They're widely talked about. They **dominate** the internet. Dominate means to rule over.

Influencers are people who set trends. They post a lot on **social media**. Social media are online communication platforms used to share information. Influencers post photos and videos. They post blogs and stories. They inspire a **following**. Following refers to fans. Fans make trends come to life. They spread trends. They make trends popular.

Some trends stick. Some trends don't. This book explores trends in technology.

Electronic technology is a big part of the modern world. Electronic refers to computers. It refers to anything that needs electricity. Electronic technology includes machines, gadgets, and the internet. The internet is a global system of computer networks. It's used to get and share information. The internet is used for communication.

Technology is also known as tech. New tech tools are being invented every day. People stand in line for hours to get the latest tech. Technology makes our lives easier.

In the 1950s, personal computers were called "electronic brains." ▶

Many people are converting their homes to smart homes.

CHAPTER 1

The Internet of Things

Being connected is trendy. We have many **smart devices**. Smart means being programmed to do things independently. Devices are tech tools and machines. For example, smart lamps turn on and off. This happens from a swipe on our smartphones. Or it happens from a voice command.

All these smart devices are connected to the internet. They're also connected to one another. They "talk" to each other. They send data. This is the internet of things (IoT). It's a giant network of connected things and people.

If a device has an on-and-off switch, then it may be a part of the IoT. Anything that can be connected will be connected. That's the future.

5G stands for fifth generation. It is considered the fastest wireless technology. This means quicker downloads. It improves video chatting. It improves gaming. Evan Fong is a gamer and YouTuber. He's known as "VanossGaming." He has over 25 million followers. He makes videos. He chats and jokes while playing games. His videos have had over 13 billion views. 4G made influencers like Fong possible. 5G will do so much more.

5G pushes the IoT further. It lets us control more devices **remotely**. Remote means from afar. Together, 5G and the IoT can create smart cities. In smart cities, everything and everyone is connected.

◄ Some say 5G is controversial. Controversial relates to causing arguments.

An autonomous car drove from San Francisco, California, to New York City, New York. It had a human driver. But the car drove 99 percent of the trip. This happened in 2015.

CHAPTER 2

Self-Driving Cars

5G makes self-driving cars possible. Self-driving cars are computer-controlled cars that drive themselves. They use 5G to connect to other cars. These cars send data to one another. They share data about speed. They share data about routes. They share data about road conditions. They share data about road dangers. They share data about traffic patterns. This helps self-driving cars move around safely.

Self-driving cars are built to sense their surroundings. They don't need humans to do this for them. Self-driving cars have cameras. These cameras see lane lines. They see speed signs. They see traffic lights.

BE A LITTLE EXTRA

SOCIAL MEDIA CHALLENGE

Your smartphone is an amazing tech tool. It can record videos of high visual and audio quality. Use it to make a music video on Triller.

Install the Triller app onto your phone.

Choose the music you want in your video. Choose the part of the song you want to feature.

Keep your phone stable. Use a tripod. Tripods are stands with three legs. They hold things.

Find good lighting. Natural light is the best. Lamps also work.

Think about your background. Make sure it's neat and simple.

Record in normal, fast, or slow motion. Zoom in or zoom out.

Do as many recordings as you need. Edit as much as you need.

Upload the video to your favorite social media sites.

Self-driving cars use different **sensors**. Sensors are tools that detect information. **Lidars** sense light. They fire out laser beams. They measure how long it takes for the beams to bounce back. They use this data to make maps. Maps are needed for directions and locations. **Radars** sense radio waves. They bounce waves around cars. They spot other cars. They spot obstacles. This helps avoid crashes.

Self-driving cars are still being developed. Elon Musk is a tech influencer. He has over 39 million Twitter followers. He owns many companies. He's building self-driving cars. He said, "Self-driving cars are the natural **extension** of active safety and obviously something we should do." Extension means an added part.

◀ Most accidents are caused by human mistakes. Self-driving cars are built to correct human errors.

Alan Turing was a mathematician. He said, "Can machines think?" His work led to AI.

CHAPTER 3

Artificial Intelligence

Self-driving cars and smart devices need **artificial intelligence** (AI) to work. AI is programming machines to think and act like humans. It's when computers can learn and solve problems without the help from humans.

For example, self-driving cars use AI. The best way to train a self-driving car is to show billions of hours of videos of real driving. These videos are used to teach the car's computer good driving behavior. This is called machine learning. A computer is given a lot of data. It learns to get better at a task. Machines can also do deep learning. Deep learning involves networks of data.

There are 3 types of AI. Weak AI is focused on doing one task very well. Examples are self-driving cars and internet search engines. Strong AI can do many things at once. An example is a robot. Super AI exceeds human intelligence. Right now, scientists have only cracked the codes for weak AI. But anything is possible!

AI is how smart devices understand human voices. It's how smart devices and apps scan faces. It's how social media uses face filters. Face filters are trendy. Ommy Akhe, also known as "autonommy," is known for a different type of face filter. She creates unique effects that work with AI technology. She re-creates **retro** Windows 95 desktops. Retro means going back in time. Her filters are funny. They're meme-worthy. A meme is a short video, picture, or phrase that goes **viral**. Viral means spread quickly.

◂ Social media networks use AI.

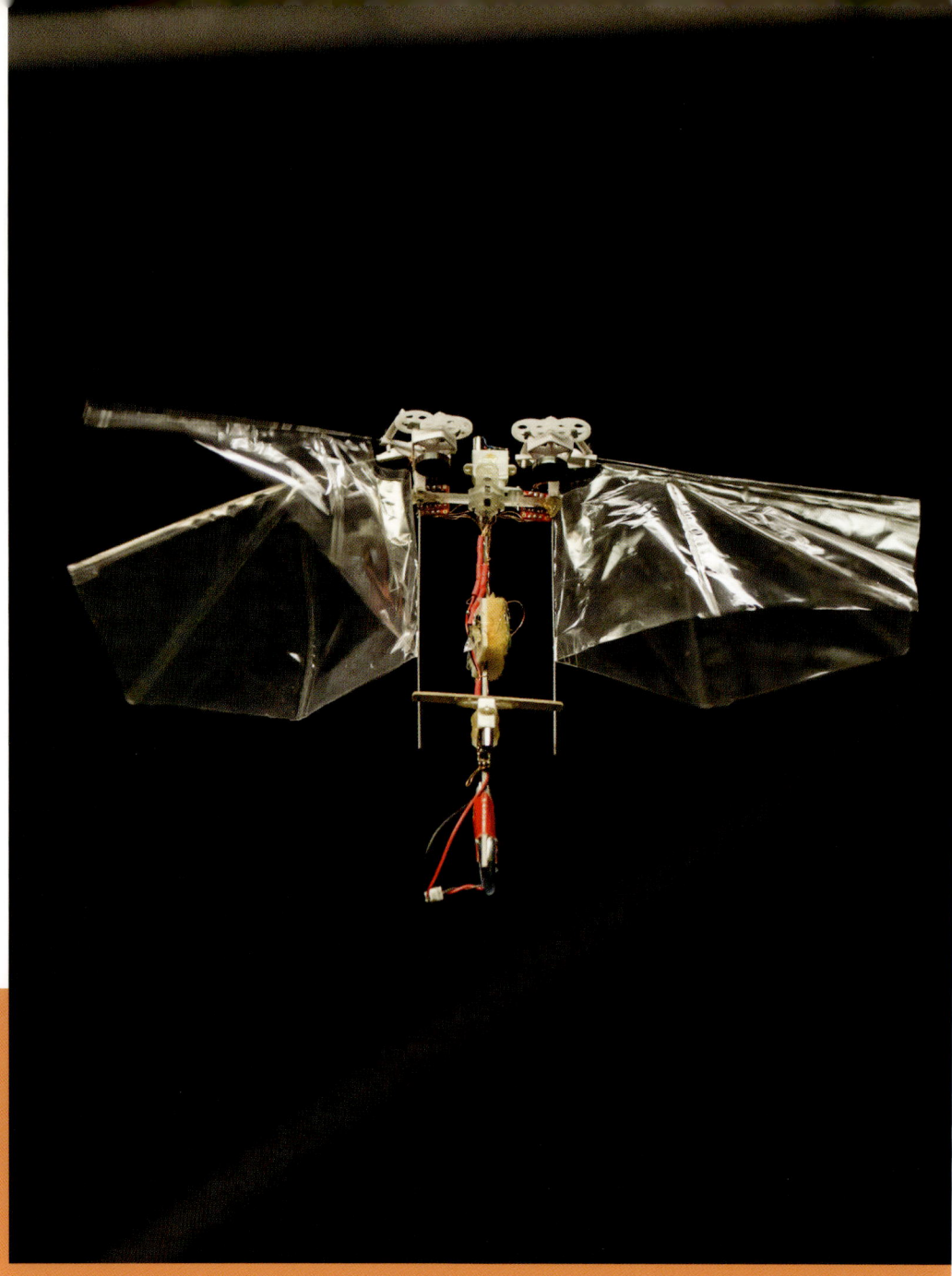

Many robots are inspired by nature and animals.

CHAPTER 4

Bionics

Robots are smart machines. They can do many tasks by themselves. They're controlled by computer programs. Technology is always improving. This means robots are getting more advanced.

Bionics is the designing of robots based on how living things move. BionicKangaroo is a robot. Scientists studied how kangaroos moved. They copied their movements. They taught the robot how to jump. The robot is smaller. It's lighter. But it moves exactly like a real kangaroo.

Bionics uses technology to improve human bodies. Scientists are working on re-creating human body parts. They've made eyes. They've made smart knees. They've made wearable kidneys. They've made bionic limbs.

Did you know...

- Steve Jobs lived from 1955 to 2011. He created the first successful personal computers. He created the iPod. He created the iPhone. He created the iPad. He focused on the look and feel of gadgets. He changed how people listened to music. He changed how people worked. He changed how people lived.

- Rayouf Alhumedhi's Twitter bio reads "Emoji creator? Maybe." She's Muslim. She wears a hijab. Hijabs are headscarves. At age 15, Alhumedhi was on social media with her friends. She wanted an emoji that looked like her. She asked big tech companies to help her. The emoji known as "Woman with Headscarf" was approved in 2017.

- Kavya Kopparapu created a deep-learning computer system. This system gives information about brain tumors. She created a computer program that uses smartphones to detect eye disease. She's the founder of Girls Computing League. She wants to influence girls to work in tech.

Bionic limbs are fake body parts. They're used to replace a missing body part. Today's bionic limbs are more lifelike. They look real. This is due to three-dimensional (3-D) printing. Bionic limbs can be controlled by the mind. They're integrated with body tissues. They're connected to the nervous system. This means bionic limbs move naturally.

Tilly Lockey is a YouTuber influencer. At 15 months old, she lost her arms and toes. At age 12, she became the first teen in Britain to get 3-D-printed bionic arms. She said, "I think of them as fashion accessories rather than just medical devices. What I love about having bionic hands is that you can have them in different colors or put lights in them, which is something people with hands can't do."

▲ Bionics combines biology and electronics.

Technology plays a big role in medical science.

CHAPTER 5

Gene Editing

Each cell in the human body has thousands of **genes**. Genes carry information that determine **traits**. Traits are features inherited from parents. Examples are eye color and hair type. Children can also inherit sicknesses from their parents.

A tech trend is gene editing. Powerful computers and tech tools can change genes. They can add genes. They can take away genes. Some scientists use gene editing to help people. They want to change genes to treat sicknesses. For example, they make people's cells stronger. This helps people fight off sicknesses.

In 2020, Helen Obando underwent gene editing, which is also known as gene therapy . She was 16 years old. She's the youngest person to do so. Doctors are still hoping that it will cure a deadly disease she has.

Some scientists use gene editing to change future generations. They get rid of genes that cause sickness. This makes sure traits are not passed down. Not everyone agrees with this. Some people fear allowing scientists to design humans.

Practicing gene editing on future generations is a hot topic. There are problems with it. Gene editing could be risky. It could be dangerous. It's also not a natural process. Many people wonder if it's the right thing to do. It seems odd to have scientists mess with genes. Some countries want to ban the gene editing of **germline** cells. Germline refers to cells that make babies.

◄ Pigs are used in gene editing experiments.

There is a sci-fi "city" at Universal Studios in Singapore.

CHAPTER 6

Sci-Fi Reality

Nerd culture. Geek chic. Being quirky is cool. Liking tech and science is cool. Science fiction, or sci-fi, combines these two things. Sci-fi refers to stories that imagine worlds with advanced science and technology. Examples are the Star Wars and Marvel Studios movies.

Many of our modern tech tools were inspired by sci-fi stories from years past. Cell phones were featured in *Star Trek: The Original Series* (1966–1969). Bionic limbs were featured in *Star Wars: Episode V* (1980). Self-driving cars were featured in an Isaac Asimov article. Asimov called them "robot-brain" cars. Smart homes were featured in Disney's *Smart House* (1999).

THROWBACK
OG INFLUENCER

There were influencers before social media. An original (or "OG") tech influencer was Countess Ada Byron Lovelace. She lived from 1815 to 1852. She was different from other women of her time. She learned math and science at a young age. At age 12, she tried to build a flying machine. She studied under Charles Babbage. Babbage was called the Father of the Computer. Lovelace added her own thoughts to a paper Babbage wrote. She described codes. She introduced many computer concepts. She knew computers could do more than just add and subtract numbers. She was the first computer programmer. She influenced the field of computer science. She encouraged women to work in this field. In 1980, the U.S. Department of Defense named a new computer language "Ada." October 15 is Ada Lovelace Day. This is a day to celebrate women's contributions to math and science. Lovelace said, "I don't wish to be without my brains."

James Hobson is known as the Hacksmith on YouTube. He studied engineering. He loves building gadgets. He's inspired by fictional ideas from comics, movies, and video games. He makes them real. He made Wolverine's claws. He made Captain America's shield. He made Iron Man's helmet and power glove. He made Thor's hammer. He films himself making these things. He uploads them to his YouTube channel. He's been a YouTube star since age 16. He has over 10 million followers.

He said, "In the future, I'd love to invent technology that improves quality of life, or enables workers to do superhuman tasks."

▲ Hobson is inspired by Tony Stark.

Glossary

artificial intelligence (ahr-tuh-FISH-uhl in-TEL-ih-juhns) the ability of a computer or computer-controlled robot to perform tasks commonly associated with human intelligence

bionics (bye-AH-niks) the science of designing robots that function like living things

devices (dih-VISE-iz) tech tools and machines

dominate (DAH-muh-nate) to rule over

electronic (ih-lek-TRAH-nik) anything related to electricity and/or computers

extension (ik-STEN-shuhn) an added part

following (FAH-loh-ing) fans or supporters

genes (JEENZ) tiny units within cells that carry information that determines your traits

germline (JURM-line) cells that make babies like eggs, sperms, and embryos

influencers (IN-floo-uhns-urz) people who set trends by posting a lot on social media and have large followings

lidars (LYE-dahrz) detection tools that use light

radars (RAY-dahrz) detection tools that use radio waves

remotely (rih-MOHT-lee) from afar

retro (RET-roh) going back in time

sensors (SEN-surz) tools that detect information using the senses

smart (SMAHRT) being programmed to do things independently

social media (SOH-shuhl MEE-dee-uh) online communications used to share information

traits (TRAYTS) features inherited from parents

trending (TREND-ing) currently a popular fad online

viral (VYE-ruhl) trending or spreading very quickly

Learn More

Discovery Kids. *How Technology Works: The Facts Visually Explained*. New York, NY: DK Publishing, 2019.

Szymanski, Jennifer. *Code This! Puzzles, Games, Challenges, and Computer Coding Concepts for the Problem Solver in You*. Washington, DC: National Geographic Children's Books, 2019.

Todaro, Joseph. *Top Secret: Crypto*. Ann Arbor, MI: Cherry Lake Publishing, 2020.

Index

Alhumedhi, Rayouf, 22
artificial intelligence (AI), 16–19
bionics, 20–23
body, human, 21, 23, 25
cars, self-driving, 12–15
computer programming, 30
deep learning, 17, 22
emojis, 22

5G, 11, 13
gamers, 11
gene editing, 25–27
"Hacksmith," 31
influencers, 5, 11, 15, 22, 23, 30
internet of things (IoT), 8–11
Jobs, Steve, 22

Kopparapu, Kavya, 22
Lockey, Tilley, 23
Lovelace, Ada Byron, 30
machine learning, 17
Musk, Elon, 15
Obando, Helen, 25
OG influencer, 30
robots, 19, 21–23

science, 6
sci-fi reality, 29–31
search engines, 19
self-driving cars, 13–15, 17, 19, 29
sensors, 15
sickness, 25, 27
smart cities, 11
smart devices, 8, 17, 19
smartphones, 9, 14, 22
social media, 5, 14, 19, 22, 30

technology, 5, 6, 21, 29
3-D printing, 23
trends, 5
Triller, 14
Twitter, 15, 22
"VanossGaming," 11
wireless technology, 11
YouTubers, 11, 23, 31

32